# My First Book of Dwarf Planets

## A KID'S GUIDE TO THE SOLAR SYSTEM'S SMALL PLANETS

### K.J. Field

PlutoShine PRESS

FOR MY FAMILY, THE CARE OF WHOM INSPIRED THIS BOOK.

**TO PLUTO AND THE DWARF PLANETS:** LEARNING ABOUT YOU OVER THE PAST SEVEN YEARS HAS BEEN A JOY. YOU ARE LITTLE MASTERPIECES. I SEE YOU SHINING THERE IN THE FARTHEST, DARKEST REACHES OF OUR SOLAR SYSTEM, AND I WILL SHARE YOUR STORY.

## Image Credits

NASA/Johns Hopkins University Applied Physics Laboratory/Southwest Research Institute: Pluto, Charon, the moons of Pluto, New Horizons (Cover, p. 1, pp 2-3, p. 15, p. 23, p. 41, p. 44, pp 46-47)
NASA/Johns Hopkins University Applied Physics Laboratory/Southwest Research Institute/Roman Tkachenko: Arrokoth (p. 2)
NASA/JPL-Caltech: Neptune (p. 10)
NASA/JPL-Caltech/SwRI/MSSS/Gerald Eichstädt/Seán Doran: Jupiter (p. 10)
NASA/JPL-Caltech/UCLA/MPS/DLR/IDA: Ceres and Dawn (Cover, p. 1, p. 30, p. 33)
NASA/JPL-Caltech/UCLA/MPS/DLR/IDA/ASI/INAF: Ceres Ahuna Mons (pp 34-35)
NASA/JPL-Caltech/ASU: Psyche (p. 2)
NASA/JPL-Caltech/R. Hurt (SSC-Caltech): Quaoar (p. 1, p. 49)
NASA, ESA, and A. Feild (STScI): Haumea (p. 8)
NASA, ESA, A. Parker and M. Buie (Southwest Research Institute), W. Grundy (Lowell Observatory), and K. Noll (NASA GSFC): Makemake & MK2 (p. 19)
NASA/JPL, ESA/ATG mediala: asteroid belt (p. 30)
Other NASA: Gonggong (Cover, p. 1, p. 15, p. 49); Hubble Space Telescope (p. 1); Sedna (Cover, p. 49); Haumea with ring (p. 6); Ida (p. 32)
Wikimedia/photo. Museum (I.Gesk) Berlin Antikensammlung: Eris mythology (p. 23)

**Visit www.theplutodiaries.com**                    **Find me on Twitter @Plutoliveshere**

# What Is a Dwarf Planet?

When you think of a **planet**, what do you imagine? Is the object in your mind round? Do you think of Earth or something Earth-like? Could you stand and walk on it? Does it have mountains and a sky?

A **dwarf planet** is a round space object that goes around the Sun. It is smaller than a gas planet or a rocky planet. A dwarf planet looks like what most people imagine when they think of a planet. In fact, it might even remind you a little bit of Earth!

So far, robot spacecraft have visited two dwarf planets! We have not yet been able to study the other dwarf planets up close. Instead, astronomers use big, powerful telescopes on Earth and in space to learn about them. In this book, all images of **Pluto** and **Ceres** are actual photographs taken by spacecraft. The pictures of the other dwarf planets are artists' illustrations based on what we know from telescopes.

**Hubble Space Telescope**

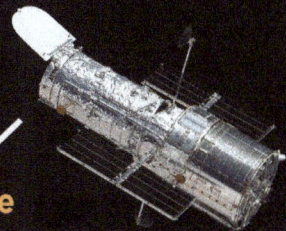

# Round Is Special

Did you know that in space round objects are unusual? Only objects that reach a certain size can be round like a ball or a globe.

**Gravity** is a force that pulls stuff together and keeps it in place. When a space object is big enough, its gravity is strong enough to pull it into a globe shape. Most objects in our solar system are too small for that to happen. This is why they are lumpy or flat. When a space object is round, it means a lot has happened inside it. It may be able to do interesting things like build mountains. Dwarf planets are round, which means they are pretty special!

Objects not to scale

**Asteroid Psyche (Illustration)**

**Dwarf Planet Pluto**

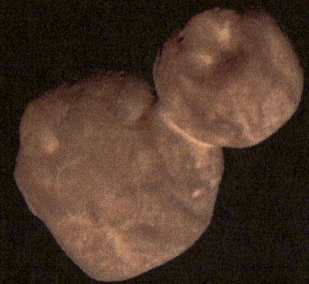

**Kuiper Belt Object Arrokoth (Photo)**

# Sharing Space

All dwarf planets except one orbit in the area of space past Neptune. We call space objects in this area **transneptunian objects**. The **Kuiper belt** is the first part of this area. Most of the dwarf planets we know about come from there. It is a gigantic place!

Dwarf planets orbit in an area with many other space objects that are about the same size. Big planets like Earth and Jupiter don't do that; they have their own space. The few other objects near them are *much smaller* than they are. The big planets have strong gravity that pushes most of the little objects away. They are too big to share their tight orbit area with other big planets.

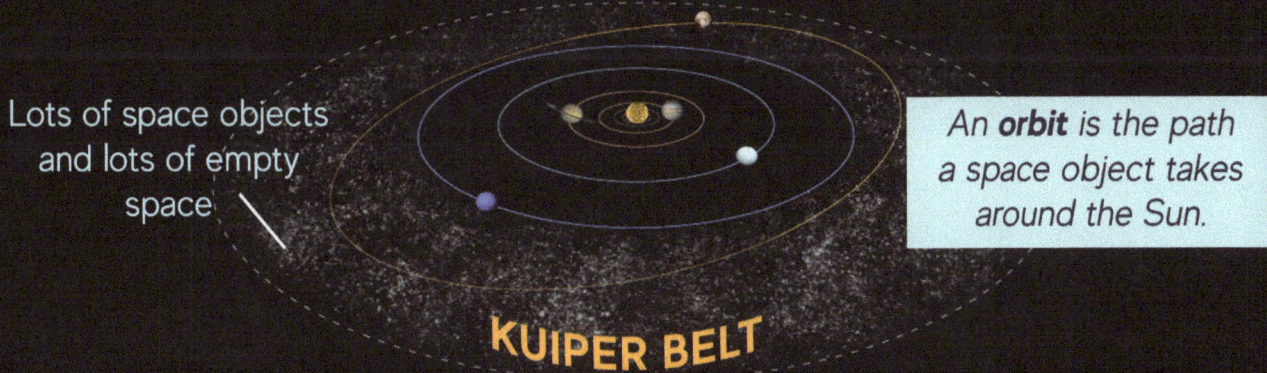

Lots of space objects and lots of empty space

An **orbit** is the path a space object takes around the Sun.

KUIPER BELT

*Haumea* is pronounced **How-MAY-uh**.

**How big?** 1,014 miles across the long way (1,632 kilometers)
If Earth were the size of a nickel, Haumea would be the size of a sesame seed.
**How far from the Sun?** 4 billion miles, or 43 astronomical units
**When did we discover it?** December 28, 2004. Brown and Ortiz discovered it.

# Haumea: the Oddball

Haumea is a special dwarf planet. It's different in many ways!

Haumea isn't round. Almost all planets and dwarf planets are shaped like **spheres**. But Haumea's shape is more like an egg or American football or rugby ball. Astronomers call that shape an **ellipsoid**. Most of the time, space objects aren't round because they're not big enough for their gravity to make them round. But Haumea is probably big enough. Something else must be going on to make Haumea oval shape.

Haumea is smaller than Pluto. Haumea's orbit sometimes brings it closer to the Sun than Pluto, but usually Haumea is farther away.

*So far away!* *Trying to see Haumea from Earth is like trying to see a coin from 100 miles away! That's about the distance from Los Angeles to Palm Springs.*

Haumea's rings are about 40 miles (70 kilometers) wide.
Saturn's rings are about 170,000 miles (273,588 kilometers) wide.

# Dwarf Planet Haumea?
# Now That Has a Nice Ring to It!

Haumea is the only dwarf planet and the only Kuiper belt object known to have rings.

Haumea's rings are hard to see, even from a telescope. We can see Saturn's rings because they are larger and the planet is much closer to Earth than Haumea is.

Scientists finally found Haumea's rings in 2017 when they watched the dwarf planet pass in front of a star. Haumea blocked the star's light for a few moments, so scientists could tell something was sticking out on both sides of the dwarf planet! Thank you, Star!

Haumea's rings are narrow and form an almost perfect circle. They spin slowly around Haumea.

*For a long time astronomers did not know smaller objects could have rings. Then they discovered that Chariklo, an asteroid-like object near Saturn, has rings. Scientists wonder why some objects have them and most don't.*

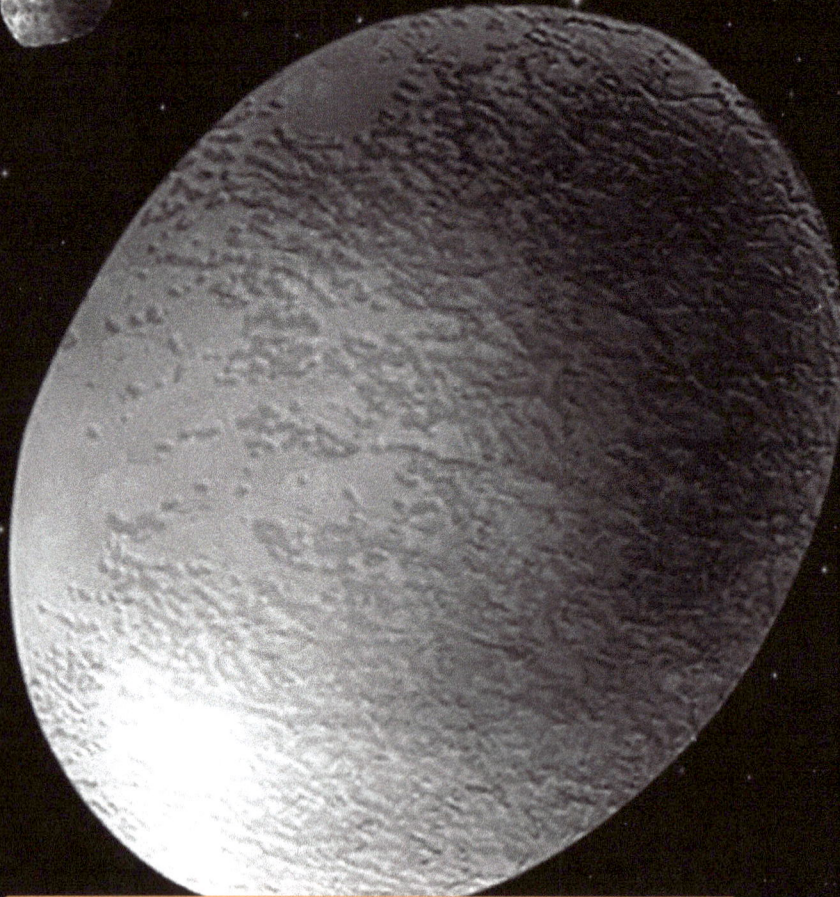

Namaka

# Three's Company

Hiʻiaka

A **system** is a group of orbiting space objects, such as a planet and its moons.

**CRASH!**

That's how Haumea got its two small moons, Hiʻiaka and Namaka: in a big **collision**! When Haumea and another space object hit each other, some pieces broke off. Haumea's gravity captured two pieces and pulled them into orbit around Haumea. Hello, moons! Haumea became a **system**.

Haumea's ring could have formed in the same crash. This probably happened about four billion years ago, in the early days of the solar system, when big crashes were common.

A few other objects in the Kuiper belt have orbits and surfaces that are similar to Haumea's. These objects may have formed in the collision, too, but Haumea's gravity didn't capture them.

*Haumea is named after a Hawaiian goddess whose two daughters were born out of different parts of her body. Sound familiar?*

An **ice mantle** is the outer covering of ice on a cold planet.
A **core** is the center of a planet.

**Do you know of any other planets that have a large spot?**

# I Spot a Spot on Haumea

Earth isn't the only planet with water. Haumea has it, too! But Haumea's water is frozen because it's so cold there. Haumea's surface is mostly made of water ice. Other dwarf planets like Pluto also have some water ice.

Haumea only has a very thin covering of ice over its rocky center. Other dwarf planets have a much thicker coating. Remember that big crash Haumea was in long ago? Astronomers think Haumea lost its **ice mantle** then. The ice coating got blasted off!

Haumea has a large, dark red spot. The spot may be a place where Haumea got hit. The red on Haumea looks like the red on Pluto. It probably got pulled up from underneath Haumea's surface when the dwarf planet was hit. The red is exciting to scientists because it's made of stuff that is important in the creation of life.

*Water ice in outer space can be different than water ice on Earth. But Haumea's water ice is like the ice in your freezer!*

# Spinning Like a Top

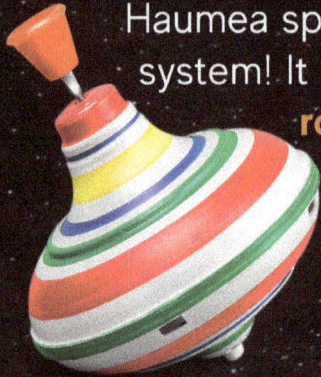

Haumea spins very fast, faster than any object in the solar system! It takes Haumea just 3 hours and 54 minutes to **rotate** all the way around. Earth takes about 24 hours. If you could live on Haumea, you'd have less than four hours each day to get everything done – and get some sleep! But you could see six sunsets and six sunrises in one 24-hour period!

Haumea's ellipsoid shape is unusual for an object of this size. Planetary scientists think Haumea is shaped like that because of how fast it's spinning. The high speed stretches and flattens it out!

*If Haumea were to spin any faster, it would stretch itself into a dumbbell shape and split in two!*

> A **day** is how long it takes a space object to **rotate**, or spin all the way around, once.

| Planet | Hours in One Day |
|--------|------------------|
| Mercury | 1,408 hours |
| Venus | 5,832 hours |
| Earth | 23 hours, 56 minutes |
| Mars | 24 hours, 37 minutes |
| Ceres | 9 hours, 4 minutes |
| Jupiter | 9 hours, 56 minutes |
| Saturn | 10 hours, 42 minutes |
| Uranus | 17 hours, 14 minutes |
| Neptune | 16 hours, 6 minutes |
| Pluto | 153 hours, 18 minutes |
| Haumea | 3 hours, 54 minutes |
| Makemake | 22 hours, 50 minutes |
| Eris | 25 hours, 54 minutes |

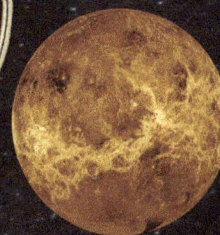

Venus

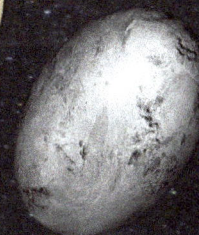

Haumea

**Makemake** *is named after the creator god of the Rapanui people of Easter Island. It's pronounced* **MAHK-ay-MAHK-ay**.

**How big?** 920 miles across (1,480 kilometers)
**How far from the Sun?** 4.2 billion miles, or 45 astronomical units
**When did we discover it?** Brown, Trujillo, and Rabinowitz discovered it on March 31, 2005.

# Makemake: the Name Everybody Loves to Say

Makemake orbits in an area of the Kuiper belt called the classical Kuiper belt. It is farther from the Sun than Pluto and Haumea. After Pluto, it is the second brightest object in the Kuiper belt. It's bright enough that you can see it from some home telescopes.

Makemake was found around the same time as Eris and Haumea. When astronomers found three objects almost as big as Pluto, some wondered whether to call them all planets or something else. Soon astronomers started using the word "dwarf planet."

Makemake is smaller than Pluto and about the same size as Haumea. It's tiny compared to Earth! Here's how the dwarf planets compare in size to the Moon.

**Makemake** and **Earth**

Pluto
Haumea
Ceres
Moon
Eris
Makemake

An **atmosphere** protects planets from the Sun's rays and from small space objects. If **oxygen** is in the atmosphere, humans can breathe it.

# The Air up There

An **atmosphere** is the gas around a planet. Astronomers can use a telescope to look for atmospheres when they watch planets pass in front of bright stars.

Astronomers couldn't find any atmosphere around Makemake, Haumea, or Eris, but they did find one around Pluto. Makemake's gravity might not be strong enough to keep an atmosphere. But it is possible that Makemake, like the other dwarf planets, may develop a thin one during the time it's in the warmer part of its orbit.

Which rocky planets have atmospheres? Earth, of course! Venus has a thick, poisonous one. Mercury doesn't have an atmosphere. Mars has a thin one.

And the giant gas planets? They're mostly atmosphere with no surface to stand on!

# Not Alone

Makemake has one **natural satellite**. It doesn't have a real name yet, but it does have a catchy nickname with numbers and letters: MK2! The little moon was discovered in 2016, eleven years after Makemake was discovered. Astronomers couldn't see MK2 for a while because it was hiding in the bright light of Makemake. It's possible that Makemake, Haumea, and Eris have even more moons that we haven't been able to see yet.

The discovery of a moon is exciting – it means astronomers can learn more about two objects! They can use the satellite (and math!) to learn things about the planet, like how big and how heavy it is. Those measurements can then help them figure out other things like what the planet is made of. MK2 helped astronomers learn some things about Makemake they wouldn't have been able to know without it.

*Not all planets have a **natural satellite**, or moon. Mercury and Venus don't have any, probably because they are so close to the Sun that its gravity steals any chance of a moon away!*

MK2's light is very dim. MK2 is 1,300 times fainter than its **host planet**!

**MK2** (photograph taken by the Hubble Space Telescope)

Makemake

A planet with a moon (or moons) is called a **host planet** or a **parent body**.

*On Earth, ice is made of water. But did you know ice can be made of other things, too?*

Like Pluto, Makemake has different kinds of gas on its surface. Makemake probably has more methane and less nitrogen than Pluto. Much of the gas is frozen because it's so cold in the Kuiper belt. Earth has these gases too, but you won't find them in their ice form there!

During the daytime, the Sun heats up the methane and nitrogen ice on Makemake's surface. Some of it unfreezes and becomes part of the air. At night, some of the gas freezes into ice again.

# Makemake, the Other Red Planet

*Mars and Makemake are both red, but not for the same reason...*

Mars is red because the iron in its rocks makes rusty-red dust. Makemake's ice looks reddish-brown because of the Sun's rays. Sunlight can turn frozen nitrogen and methane into sticky, dark material that covers the ice. This happens on Pluto, too.

Makemake has small grains of methane ice on its surface. The size of the grains is about one centimeter, or about as wide as an adult pinkie finger. Isn't it amazing that scientists can figure out these grains exist without seeing Makemake's surface?

*Eris* is pronounced **EER-uhss** or **AIR-uhss**.

**How big?** If Earth were the size of a nickel, Eris would be the size of a popcorn kernel.

**How far from the Sun?** 6 billion miles, or 68 astronomical units

**When did we discover it?** Brown, Trujillo, and Rabinowitz discovered it on January 5, 2005.

# Eris: Troublemaker

Dwarf planet Eris is named after a Greek goddess who was a real troublemaker! She caused arguments everywhere she went. The dwarf planet, too, caused a lot of trouble when it was discovered. At first, astronomers thought Eris was bigger than Pluto. That made them argue over whether to call Pluto a planet!

Eris is actually the second biggest dwarf planet by **diameter**. Pluto is the biggest. But Pluto and Eris are very close in size.

Eris is smaller but more **massive** than Pluto. This means it weighs more. Eris is heavier probably because it has more rock and less ice.

*Diameter is the measurement across the center of a circle.*

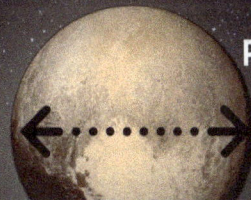

Pluto

Eris

1,477 miles across
(2,377 kilometers)

1,445 miles across
(2,326 kilometers)

# Far and Away

Eris is the farthest of the five dwarf planets. It orbits in an area past the Kuiper belt called the **scattered disc**. Eris is the same distance away as traveling from the Sun to Pluto three times!

The amount of time it takes a space object to go all the way around the Sun once is called a **year**. Planets that orbit farther away have a longer path to travel, so their year is longer.

Mercury is the closest planet to the Sun. It only takes Mercury 88 days (in Earth time) to go around the Sun once. It takes Eris about 557 years! That means one year on Eris lasts 557 Earth years. If you could live on Eris, you wouldn't even make it to your first birthday!

| Dwarf Planets: Length of a Year in Earth Time | | |
|---|---|---|
| **Earth:** 1 year (365 days) | **Pluto:** 248 years | **Makemake:** 310 years |
| **Ceres:** 4 years, 7 months | **Haumea:** 283 years | **Eris:** 557 years |

The coldest temperature recorded on Earth is -128.6 degrees Fahrenheit. It was recorded in Antarctica on July 21, 1983.

It's very cold on Eris! The surface temperature depends on where Eris is in its orbit.

**Farthest from the Sun:** -406 degrees Fahrenheit (-243° Celsius)

**Closest to the Sun:** -359 degrees Fahrenheit (-217° Celsius)

# Solar System Orbits

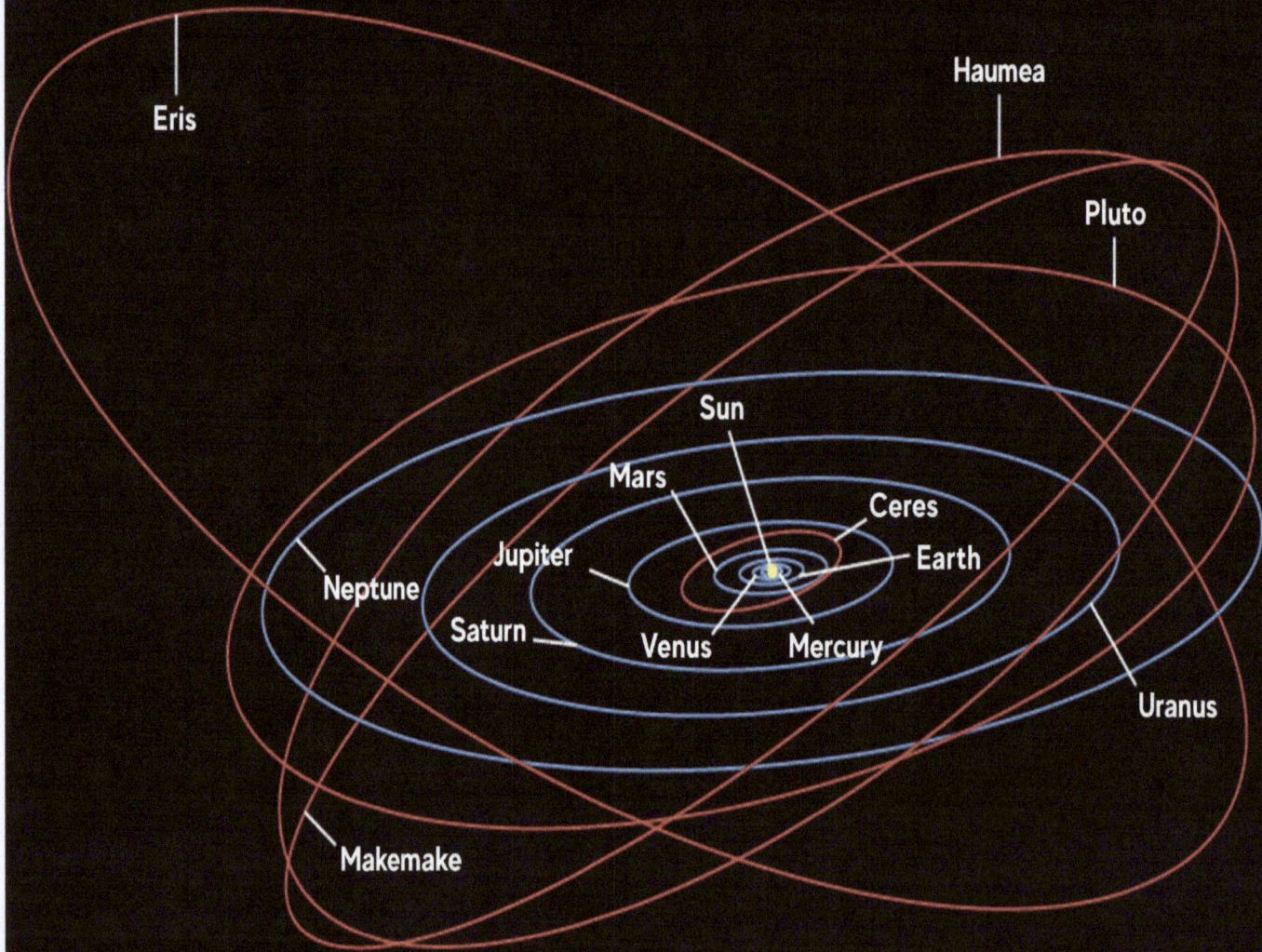

# Epic Orbits

Eris has a weird and wonderful orbit! It's long and stretched out like a squashed oval. The oval shape means the dwarf planet is farther away from the Sun during parts of its orbit and closer to the Sun during other parts. Eris' orbit is also quite tilted. Astronomers had a tough time discovering Eris because its orbit is so tilted. Eris was hard to see, and they weren't looking in the right places!

The rocky and gas planets have mostly circular and flat orbits. Dwarf planets in the outer solar system can have different kinds of orbits: circular, oval, flat, and tilted.

Dwarf planets' orbits cross over each other. But the bigger planets can't have orbits like that – they have to travel in their own lanes! The outer solar system is different than the inner solar system. It is so big that crisscrossing orbits isn't a problem – no body bumps into each other!

*An **inclined** orbit is an orbit that looks tilted.*
*An **eccentric** orbit is an orbit that is oval-shaped instead of round.*
*Some orbits are more inclined and eccentric than others.*

# Like Yin...

Eris' surface is covered in ice. It looks almost white and is as bright as fresh snow. Eris has mostly nitrogen and methane ice like Pluto and Makemake. Scientists know this because powerful telescopes can see the unique light each chemical gives off.

When the Sun shines on a space object, some of the sunlight bounces off the surface of the object. Eris is special because it's very **reflective**. This means almost all the sunlight that hits the surface bounces off. It makes Eris look bright. The only space object in the solar system that is more reflective than Eris is Saturn's shiny moon Enceladus.

*It takes sunlight **9 hours** to travel from the Sun to the surface of Eris. It takes about **8 minutes** for sunlight to reach Earth!*

# ...and Yang

Eris has one moon named Dysnomia (dice-NOH-mee-uh).

Dysnomia is big for a moon. Its large size means it's probably round like a planet. It is the second biggest transneptunian moon after Pluto's moon Charon. Astronomers used Dysnomia's size and orbit path to figure out that Eris is heavier than Pluto.

Dysnomia looks very different from Eris. While Eris is bright white, Dysnomia is very dark. It might be as black as coal. Eris and its moon are as different as night and day!

*Dysnomia is 435 miles across (700 kilometers). That's bigger than Saturn's famous moon Enceladus.*

Unlike Earth and Pluto, Ceres doesn't have seasons. This is because it spins upright and doesn't tilt.

**How big?** 580 miles across (940 km).
It has the same amount of land (surface area) as the country of Argentina.
**How far from the Sun?** 257 million miles (414 million kilometers), or about 3 astronomical units

# Ceres: Queen of the Asteroid Belt

Ceres is unique. It is the only dwarf planet without a satellite and the only one in the inner solar system. It orbits in the asteroid belt, a ring of space between Mars and Jupiter. The asteroid belt has lots of small objects, but it is still mostly empty space.

Ceres is the smallest dwarf planet, but it's a giant in the asteroid belt! It's much bigger than any of the asteroids. Ceres is also the only true sphere in the belt.

Ceres is warmer than other dwarf planets because it's closer to the Sun, but it is still much colder than Earth. Its surface is a mixture of water ice and minerals such as carbon and clay. Ceres spins almost completely upright. This is different from Earth, which tilts a little as it spins.

*Ceres is named after the Roman goddess of harvest and grains. So is the word "cereal"!*

# More Than an Asteroid

Ceres was the first dwarf planet and the first member of the asteroid belt to be found. It was discovered on January 1, 1801 by astronomer Giuseppe Piazzi. Piazzi thought there might be a regular-size planet in the space between Mars and Jupiter. He found Ceres instead.

Back then, nobody knew about dwarf planets. At first, people called Ceres a planet. Later they called it an asteroid when they realized how small it is. Many astronomers used both words to describe Ceres for a long time.

As telescopes got better, astronomers saw that asteroids are very different from planets. Asteroids aren't just smaller than planets; they're lumpy, too. That's because their gravity is too weak to mold them into a ball shape. But Ceres is not like that. It's round and similar to rocky planets like Mars in many ways.

**Asteroid Ida**
(NASA photo)

In 2015, the NASA spacecraft **Dawn** went into orbit around Ceres. It spent several years taking photos and sending us data.

**Ceres** is pronounced **SEER-eez**.

# Bumps, Bruises, and Mystery Spots

Ceres has many **impact craters**. Some were created billions of years ago. One special crater on Ceres is Occator crater.

The Dawn spacecraft discovered hundreds of very bright spots on Ceres' surface. The largest bright spot is in Occator crater. This spot is called Cerealia Facula.

The twinkling spots are one of Ceres' mysteries, but scientists think they are made of a kind of salt. Ceres may be hiding a frozen ocean of salty water underground! Salt from the ocean seems to be passing through to the surface creating those bright spots. It is possible the salty water could provide a home for life.

*Occator crater* is 50 miles across (80 kilometers).

Cerealia Facula

Ceres has more water than any place in the inner solar system except Earth!

An **impact crater** is a bowl-shaped hole in the ground where a space rock hit.

Ceres sometimes has a very thin atmosphere made of water **vapor**, or mist. When the Sun heats Ceres' surface ice, some of it turns into water vapor.

# More Than a Mountain

The spacecraft Dawn took photos of one of Ceres' most fascinating features: Ahuna Mons. It's a tall mountain with bright streaks running down it. Ahuna Mons is big for such a little world! Planetary scientists think Ahuna Mons isn't *just* a mountain – it's a volcano! Very cold planets like Ceres can have a special kind of volcano called a **cryovolcano**, or ice volcano. Ceres' cryovolcano erupts a mixture of salty water and mud instead of hot lava.

If it weren't for Jupiter's strong gravity right next door, Ceres might have formed into a larger planet billions of years ago.

Ahuna Mons is young, for a volcano that is! It may be only a few hundred million years old. Scientists think that a new cryovolcano bursts out of Ceres' ground every 50 million years. Right now, Ahuna Mons is the only ice volcano on Ceres, so something must have happened to the other volcanoes. Over time, they may have flattened out and settled back into the ground. This can't happen to Earth's volcanoes because they're made of rock.

*Ahuna Mons is 2 1/2 miles high (4 kilometers). This 3D image was created by combining many different pieces of information from Dawn.*

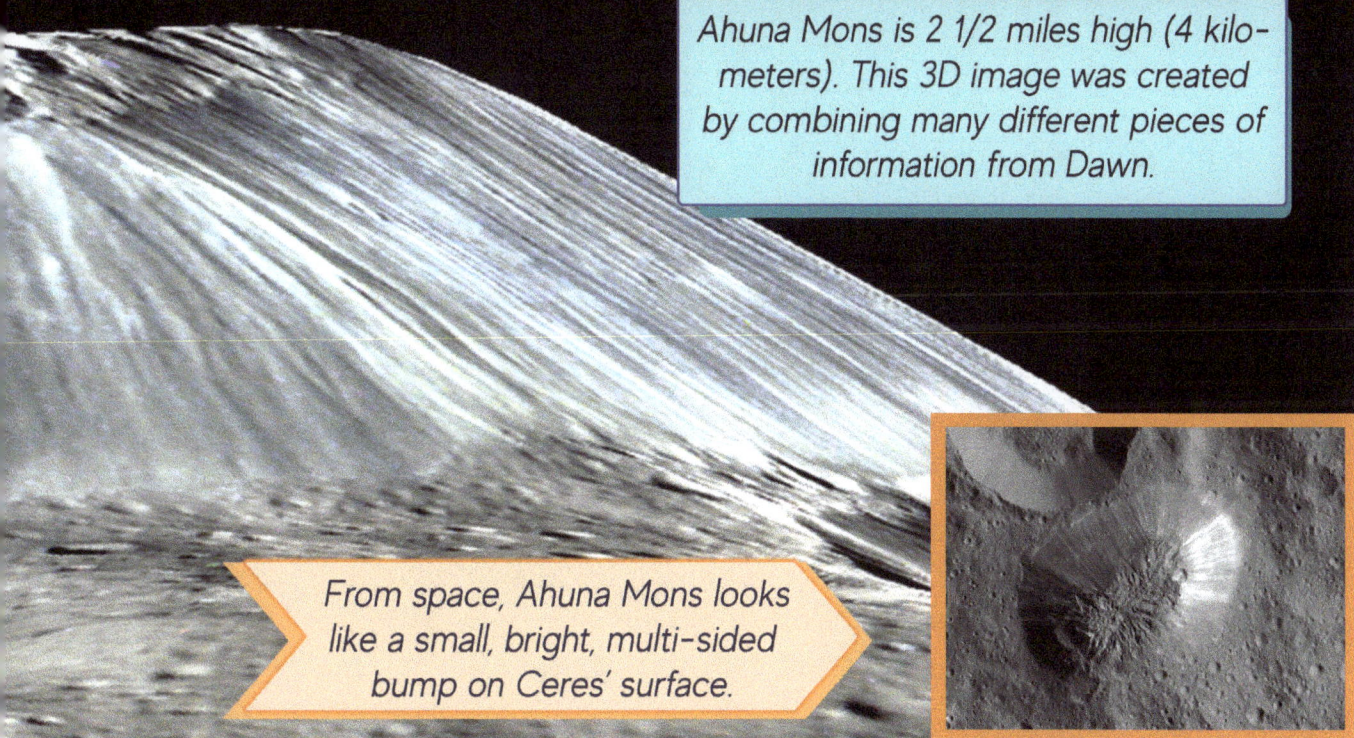

*From space, Ahuna Mons looks like a small, bright, multi-sided bump on Ceres' surface.*

*Pluto is about half as wide as the United States of America.*

**How big?** 1,477 miles across (2,377 kilometers)
**How far from the Sun?** 3 1/2 billion miles, or 39 astronomical units
**When did we discover it?** Clyde Tombaugh discovered it on February 18, 1930 at the Lowell Observatory in Arizona.

# Dwarf Planet Pluto Is Kind of a Big Deal

There are many things about Pluto that make it interesting and different! Pluto spins in the opposite direction as most other planets. It's also tipped on its side.

Pluto seems small when you compare it to rocky planets and gas planets. In fact, seven moons are bigger than Pluto! But Pluto is still big when you think of all the even smaller objects in the solar system. There are billions, and even trillions, of those little objects! Pluto is the largest dwarf planet and the largest known transneptunian object.

Pluto is *big* for another reason. It's the farthest planet (and the second farthest object) that's ever been visited by a spacecraft. And it surprised astronomers in a very *big* way!

*Including the Sun and satellites, Pluto is the 17th largest object in the entire solar system! Not so little when you think of the big picture...*

# Family Portrait

Pluto has five moons. The biggest is **Charon** (SHAIR-uhn or SHAR-uhn). Charon is half Pluto's width and is pretty heavy, even compared to Pluto. That is very large for a satellite! Some astronomers treat Pluto and Charon like two planets of one system. They call them a **double planet**. In some ways, Charon doesn't act like a moon. A moon orbits the larger object. But Charon does not exactly orbit Pluto. The two objects look like they orbit each other as they go around the Sun!

Charon has a long, deep **fault** across its surface. This is where the ground has pulled apart. It's like a belt that wraps around Charon. It's the longest in the entire solar system! Charon also has a large dark spot like Haumea.

The other moons are very small. Styx is only 10 miles (16 kilometers) across the longest way! If you could walk across Styx with no problems, it would only take you about four hours. The little moons orbit both Pluto and Charon. They spin and tumble very fast around them. Hydra and Kerberos look like they are made of several objects put together.

*Pluto's moons may have formed in the same way as Haumea's moons: in a crash between two objects!*

Charon's group of **faults** is as long and as deep as four of Earth's **Grand Canyon** put together!

Faults

Grand Canyon, Arizona

## Charon and the Small Moons of Pluto

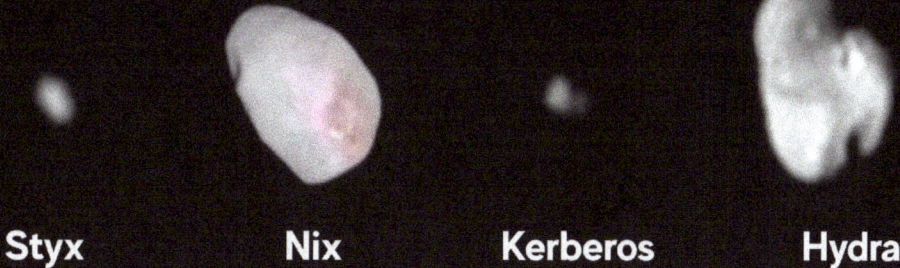

**Styx**

**Nix**

**Kerberos**

**Hydra**

Scientists think the nitrogen ice in Pluto's heart may be soft like toothpaste!

**New Horizons** is a robotic spacecraft that traveled more than nine years and three billion miles to reach Pluto. It flew by Pluto on July 14, 2015 and took some amazing photos.

# Big Heart, Big Personality

For a small planet, Pluto has some gigantic places! When the spacecraft **New Horizons** flew past Pluto, NASA was amazed to see a huge heart-shaped region covered in bright nitrogen ice. The heart is 1,000 miles wide, about the size of Texas and Oklaholma put together. It's so big New Horizons could see it shining in the distance from 70 million miles away!

Pluto's heart is called **Tombaugh Regio** and is named after Pluto's discoverer, Clyde Tombaugh. Planetary scientists think half of the heart may actually be a deep, bowl-shaped crater that has filled in with ice. Pluto's heart has many interesting patterns and shapes on it. Some of the dots you see on the smooth side are floating ice hills.

Pluto showed scientists that it is not a dead world. Underneath its icy heart, the planet is probably still warm. Pluto's heart and other areas are still changing in size, shape, and color.

# Like Earth with a Twist

When we finally saw Pluto, it was a big surprise. Pluto is like a strange alien world with some funny things that remind us of our planet.

Pluto has an atmosphere. It has blue skies just like Earth – but the sky is darker. It has clouds, wind, and weather. It has something that falls from the sky like snow, only it's made of methane – and it's reddish brown! It has icebergs, tall mountains, and two cryovolcanoes. It has high, hilly places and low, flat places. It has faults, valleys, cliffs, pits, sand-like dunes, and even a frozen lake made of nitrogen. There is a huge ocean of water, too, but it's underground, below Pluto's heart! Nobody knew such a little world in such a far-off, cold and dark place could have so much going on.

*From Pluto, the Sun looks bright but very small, like the stars we see at night. Even though the Sun is far away, it still gives Pluto heat and light.*

In 2006, a group of astronomers said dwarf planets and planets should be separate kinds of space objects. They said a planet is something that does **not** share its orbit with other objects its size.

But many planetary scientists still think of Pluto as a small type of planet. Why? Because they are not thinking about orbits. They are thinking about the things on and around a dwarf planet, like mountains and atmosphere. In these things, dwarf planets are similar to the rocky planets Earth, Mars, Mercury, and Venus. In fact, the rocky planets have more in common with Pluto than they do with giant gas planets.

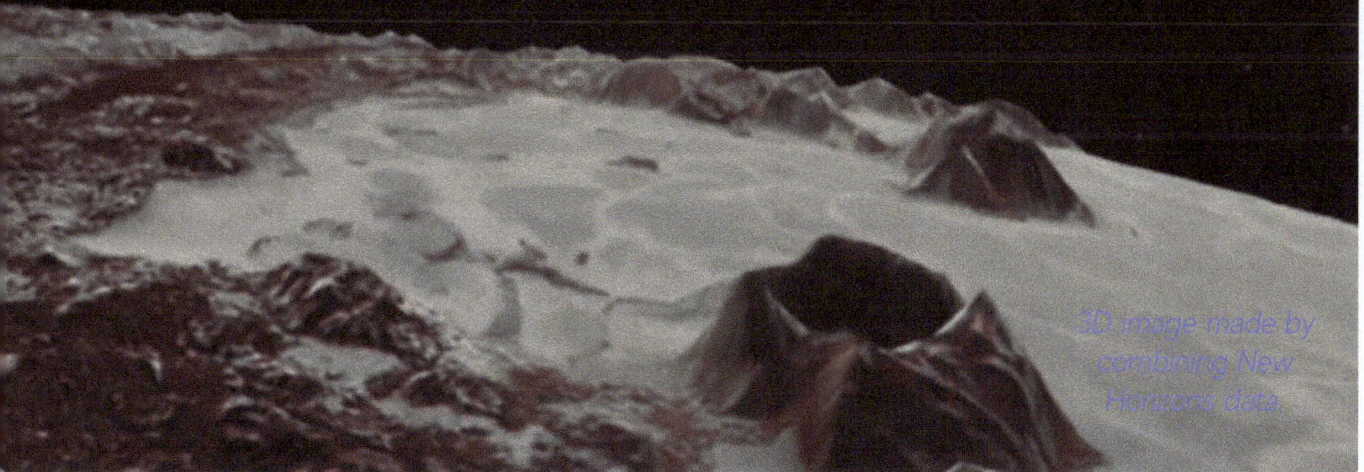

3D image made by combining New Horizons data.

# Just Five? The Other Dwarfs...

There are lots more dwarf planets than the famous five! Astronomers have already discovered many objects in the region beyond Neptune. All of the round ones are dwarf planets. These objects are very far away, so it's hard to tell if they are round, even with a telescope. But astronomers can make good guesses by figuring out their size and **composition**.

There are actually hundreds of dwarf planets in our solar system. Hundreds?! Does that number seem big? It's not so big when you realize there are millions upon millions of space objects out there. Soon we will learn about dwarf planets in other solar systems, too.

The space objects on the opposite page are some of the biggest dwarf planets that have been found. They were all discovered by astronomer Mike Brown and his team. We have no close-up photos of these dwarf planets yet, so the pictures are artists' illustrations based on science.

*A space object's **composition** is what it is made of. Dwarf planets **are composed of** rock and ice, but differ in the amount and kinds of ice they have.*

## GONGGONG

**Size:** 760 miles across (1,230 kilometers). Slightly smaller than Makemake.
**Moons:** 1 **(Xiangliu)**
**Fact:** One of the reddest objects in the solar system.

## QUAOAR (KWA-war)

**Size:** 697 miles across (1,121 kilometers).
**Moons:** 1 **(Weywot)**
**Fact:** Has a small amount of methane on its surface. In 2023, two narrow rings were discovered.

## SEDNA

**Size:** 618 miles across (995 kilometers).
**Moons:** None discovered.
**Fact:** Orbits in the Oort Cloud. One of the farthest known dwarf planets in the solar system.

## ORCUS

**Size:** 570 miles across (910 kilometers). About the same size as Ceres.
**Moons:** 1 **(Vanth)**
**Fact:** Has a lot of water ice and may still have ice volcanoes.

If you enjoyed this book, would you kindly consider leaving a review? Reviews are tremendously important to an author's visibility in a crowded marketplace and help other readers find their next book. I look forward to reading your thoughts. Thank you!

## About the Author

KJ Field is an author, screenwriter, and space enthusiast who's been writing about the solar system since 2015. She has authored a number of space-themed fiction and nonfiction children's books, with many more on the way! KJ Field has been featured on BBC Radio Five Live for her expertise on dwarf planets. She is the voice of the Planet Pluto on X (Follow her @Plutoliveshere) and hopes to someday soon turn her screenplay about the plights and delights of a personified Pluto into an animated movie for all ages. To learn about new releases and special offers, please join her mailing list at theplutodiaries.com/plutoshinepress

## Free Dwarf Planets Activity Pack

Sign up for my mailing list and receive a free printable activity pack filled with word searches, crossword puzzles, and coloring images specially created to supplement the material in My First Book of Dwarf Planets!

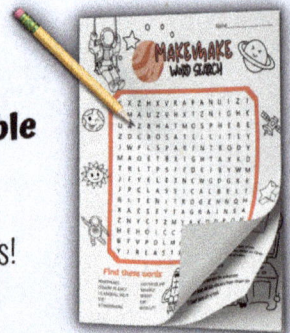

# Other Books by KJ Field

Check out PlutoShine Press' growing catalogue of children's space books, available at online bookstores everywhere!

**5 Little Dwarf Planets:** *A playful, science-packed introduction to Pluto and the dwarf planets told in sweet rhyming couplets. Introduce the solar system's five dwarf planets to children ages 3-8 with this fun planetary tale that proves small planets can shine just as bright as big ones..*

**Jupiter the Gassy Giant:** *Jupiter loves his gassy surface. If only every body felt the same way! Moons Io and Ganymede are in for some aromatic surprises as they take a whiff of the Solar System's Gassiest Giant. But can two moons who don't know the joys of a gassy surface learn to appreciate the pungent aroma of their ginormous host planet? Take a deep dive into the scents of Jupiter with this adorable picture book as you enjoy a unique story approach to children's astronomy and chemistry. Ages 5-10.*

**The Night Luna's Light Went Out:** (fiction) *When self-conscious Luna realizes she can never compare with magnificent Earth, she puts her moonbeams away for good. But the consequences are far more severe than she could ever have imagined! Will Luna get her light back before it's too late? A sciencey tale of cosmic friendship that proves every body in our solar system has a special and important role to play. Ages 4-9.*

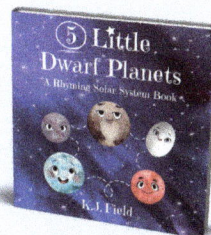

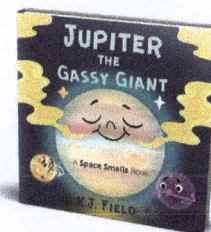

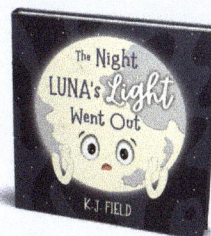

www.ingramcontent.com/pod-product-compliance
Lightning Source LLC
Chambersburg PA
CBHW052345210326

41597CB00037B/6267

9 781955 815031